The State Capitol of California, that is to endure for generations, should be a structure that the future will be proud of, and surrounded with a beauty and luxuriousness that no other Capitol in the country could boast.

Governor Leland Stanford,
Annual Message,
1863

California State Capitol Restoration

514261

Written and compiled by Lynn G. Marlowe

Edited by John C. Worsley, FAIA and Dale E. Dwyer, AIA

Published by the California State Legislature,
Joint Committee on Rules

Library of Congress Catalog Card Number 83-620000

ISBN 0-9611168-0-3

Printed in the United States of America

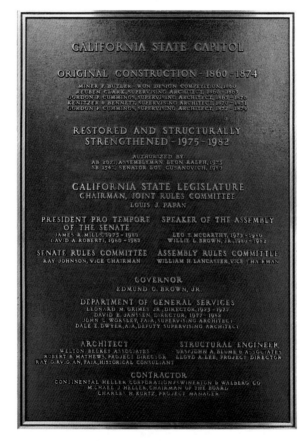

This bronze plaque commemorates the history of the
California State Capitol and was installed during the
Capitol Restoration Gala in January, 1982.

TABLE OF CONTENTS

California's Legislature moved into its new State Capitol in 1869. By the 1890s, the building's decorative work was perfect. Grand walnut stairs rose from marble and ceramic floors; two-story legislative chambers were reminiscent of Britain's Parliament. The 120-foot inner dome displayed the finest examples of Victorian decoration such as fleur-de-lis designs, cast iron bear heads, and golden spheres. The exterior was designed in a classic Roman Corinthian style.

Time and a booming population proved severe on the building. Frequent remodelings helped to accommodate a growing government but disturbed the original interior. Staircases were removed, chandeliers taken down, and high ceilings divided by mezzanine levels. By 1972, the Capitol had lost its original architectural integrity and could no longer function for its original purpose. Finally, a seismic study declared the building unsafe in a major earthquake.

In 1975, the decision was made to restore the Capitol to its turn-of-the century elegance instead of constructing a new building. Incompatible needs of structural strengthening, restoration, and 20th century modernization were to be reconciled. Nothing quite like the California State Capitol Restoration had been attempted before.

Restoration is not uncommon, yet the Capitol Project was a once-in-a-lifetime opportunity for those involved. Care was used in taking the building apart; research, dedication, and love helped put it back together. Today, California's Capitol functions as a strong, modern building while retaining its historical ambience. The Capitol is still the Legislature's home. It is a moment of the past saved for the future.

HISTORICAL

BACKGROUND. Construction work began on the Capitol in 1860. Fires, floods, and the Civil War all slowed the building's progress. After fourteen years, the Capitol was completed.

The Capitol's Roman Corinthian style featured cast iron ornamentation. At first, the smaller pieces arrived to the Capitol site on wagons, pulled by mule teams. The enormous portico and colonnade columns followed. Each column was thirty feet long, four feet in diameter, and weighed eleven and a half tons. The cast iron walls were four inches thick. Arriving either by water or rail, the columns were hauled from Sacramento's waterfront to the site by a steam tractor *(This page; bottom)*. This modern machine of 1871 moved at a rate of approximately one mile per hour.

The original plans called for an iron fence to enclose the grounds. Instead, this picket fence *(Opposite page; top left)* was constructed from leftover scaffolding and remained until 1880 when the iron fence was finally built.

The east apse *(Right)*, a semi-circular structure, originally housed the State Library and the Supreme Court.

The Capitol was overcrowded by 1949. More office space was desperately needed. Instead of constructing additional space near the Capitol, the historic apse was razed *(Center left)* and the new, six-floor annex built *(Bottom left)*.

A major remodeling project in 1906 was the beginning of many changes which altered the original design of the Capitol. By the 1970s, cubicles, divided doorways and divided windows *(This page; bottom)* illustrated how little remained of the building's Victorian dignity.

The State Architect's 1972 report declared the Capitol unsafe in a major earthquake. The original brick used in constructing the Capitol was not reinforced with concrete and steel, as buildings are today. The first solution was to abandon the old Capitol and build a larger, modern one. But, by 1975, a renewed pride in California's heritage led to another idea: restoration. The California State Capitol was given another chance.

(Opposite page) Structural strengthening was needed to make the Capitol a safe home for California's legislature. Aesthetically, the building would be restored to the years 1900-1910, a time of architectural integrity and grandeur.

DESIGN PLANS. A copy of the 1978 group picture was signed by the workers and enclosed in the cornerstone's time capsule.

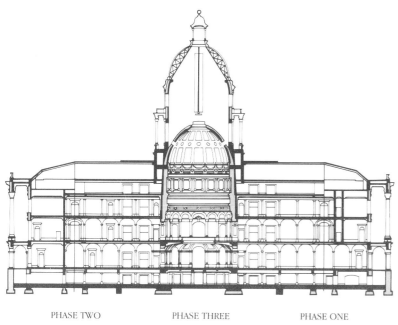

PHASE TWO   PHASE THREE   PHASE ONE

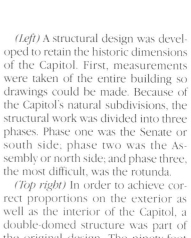

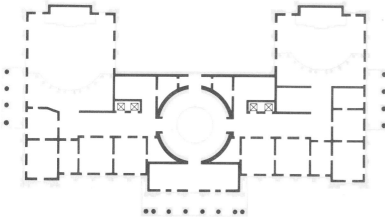

(Left) A structural design was developed to retain the historic dimensions of the Capitol. First, measurements were taken of the entire building so drawings could be made. Because of the Capitol's natural subdivisions, the structural work was divided into three phases. Phase one was the Senate or south side; phase two was the Assembly or north side; and phase three, the most difficult, was the rotunda.

(Top right) In order to achieve correct proportions on the exterior as well as the interior of the Capitol, a double-domed structure was part of the original design. The ninety-foot space between the exterior copper dome and the interior decorative dome contains a spiral stairway *(Bottom right)* leading to the cupola.

INVESTIGATIVE

WORK led to an understanding of the original construction *(This page).* During this stage, workers discovered pieces of original frieze work, lincrusta, rosettes, and other fragments. Molds were cast to record the pieces.

(Opposite page) Original woodwork was removed piece by piece from the Capitol and taken to a warehouse for stripping and refinishing. A catalog number identified each piece for reinstallation.

STRUCTURAL

SOLUTIONS. Vertical and lateral supports, called shoring and bracing, temporarily supported the original exterior walls during demolition and reconstruction of the interior.

Structural work began with phase one on the south wing. The first step was to remove the 1906 roof. Its primitive construction consisted of steel straps spaced two feet apart for reinforcement *(Top left)*.

Working one floor at a time from the fourth floor down, twelve inches of original brick was removed from the inside face of the exterior walls.

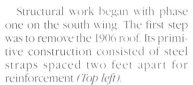

Once the layer of original brick was removed from the walls, holes were drilled into the remaining brick *(Bottom left)*. New steel anchors were secured in the holes with epoxy. Two layers of steel reinforcing bars were tied to these anchors.

To restore the walls to their original thickness, a wet-process gunite called shotcrete was blown from a nozzle over the steel reinforcing bars *(Top)*. The cured shotcrete formed twelve inches of new, reinforced concrete to carry the old brick walls. The method of spraying shotcrete allowed workers to reinforce the walls gradually, from the top down. Removing a twelve-inch layer from the entire height of the wall would have left the structure weak and unsupported.

Even though the original basement foundations had settled during their one hundred years of use, the foundations could support both old and new structures when the old brick was cored and new needle beams inserted *(Top)*. The result was a new three-foot thick concrete mat foundation tied to the existing foundations and exterior walls *(Bottom)*.

A new steel roof truss system was installed at the roof level so the fourth floor could be suspended below it.

Construction methods for phase two on the north wing were almost identical to those used on the south wing.

Once the basement foundations had cured, the new interior walls and floors were poured and tied into the reinforced exterior walls *(Left)*.

A temporary bracing frame *(Top right)* held up the cast iron columns during the undermining and reinforcing stage. The exterior porticos were reinforced by installing a new foundation under each pier at the basement level.

Next, the granite piers were hollowed out at the first floor level *(Bottom right)* so a new reinforced concrete column could be poured through the old exterior granite and the cast iron column.

The rotunda reinforcement in phase three called for a variation on the plan used for the two wings of the building. This difficult phase involved ten major steps as shown in the diagram.

◆ *One.* The new first and second floors were installed on a temporary foundation.

◆ *Two.* Scaffolding was erected in the entire area.

◆ *Three.* Concrete needle beams were installed through the brick.

◆ *Four.* The area under the dome was gunited.

◆ *Five.* The upper drum wall was gunited.

◆ *Six.* A temporary bracing was installed so the cast iron colonnade could be hung below it.

◆ *Seven.* A layer of brick was removed under the colonnade.

◆ *Eight.* The drum wall was wrapped with gunite from the top to the basement.

◆ *Nine.* A new foundation was installed at the basement level.

◆ *Ten.* New walls and floors were conventionally poured.

Reinforcing anchors were set in the brick of the inner dome *(Top right)* to prepare for step number four, guniting of the under dome. Reinforcing bars were tied to the anchors prior to placing the gunite shell.

Temporary steel columns *(Center right)* were part of the shoring and bracing system in step number six, reinforcing of the colonnade.

The entire cast iron colonnade was supported by these temporary columns during reinforcement *(Bottom right)*.

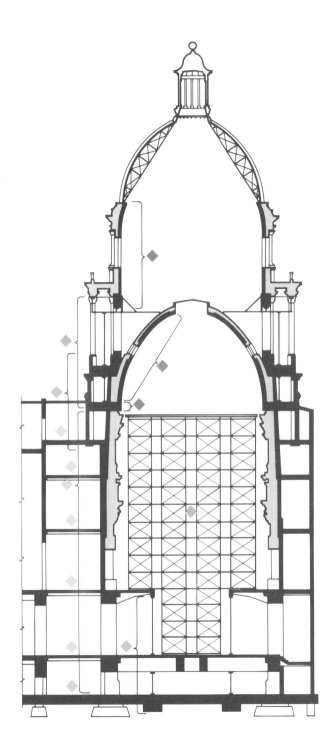

All twenty-four brick piers between the colonnade windows were removed two piers at a time and reinforced with concrete *(Top left)*.

Cast iron pilaster caps and window frames were held in place by the temporary steel bracing *(Top center)*.

The drum wall was wrapped with gunite from the top to the basement *(Top right and bottom)*, as illustrated in step number eight on the opposite page.

EXTERIOR RES-
TORATION. Various colors decorated the Capitol's exterior copper dome over the years. The original shingle system was never efficient; the dome always leaked during Sacramento's rainy winters. A newly-created system solved the leaking problem. Duplicates of the original shingles were rolled and locked together with new, waterproof ribs. The original copper ribs were restored and placed over the new ones.

(Bottom left) A life-size mock-up of a portion of the dome was constructed to test the new system.

Approval of the new system meant that the old copper could be removed from the dome and the new copper applied. Stainless steel clips *(Top right)* attached the original ribs to the new ones.

The restored dome *(Bottom)* has an untreated copper finish which will oxidize slowly to appear as it did prior to 1900.

The original gold ball was placed on the top of a new mast *(This page; left)*. The ball had been gold plated over copper in 1871 with three hundred dollars of melted gold coins contributed by the public.

New gold leaf refinished the cupola roof in 1953 but was badly deteriorated by the time of the restoration *(Top right)*. An electroplated gold finish replaced this earlier work *(Bottom right)*.

Exposure to the elements had weakened the building's hold on the cast iron pieces; corrosion had deteriorated the bolts. The ornaments were removed from the building *(Opposite page; top left and bottom left)*, and each piece was given an identification number. Sandblasting *(Top right)* removed old lead paint so that an anti-rust primer could be applied *(Center right)*. Then, the pieces were hung to dry *(Bottom right)*. Missing iron pieces were recast to match the original ornaments. New stainless steel anchor bolts hold the iron permanently on the building.

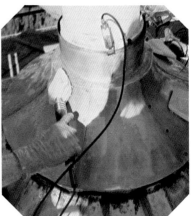

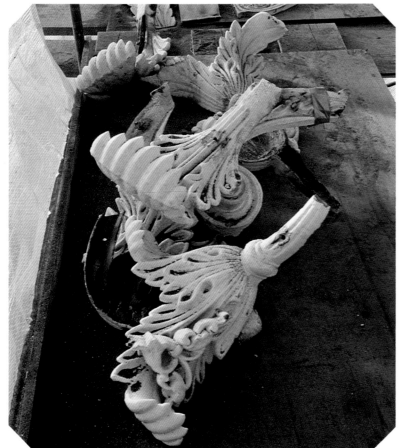

When the fourth floor was added to the building in 1906, a parapet, or solid wall, was built around the roof. Originally, the roof had a Ransome stone balustrade, or fence-like border. The balustrade was recreated out of concrete. Fragments of the original and historic photographs served as a guide for the new balustrade *(Top left)*.

Statuary decorated the Capitol's balustrade until 1906. At that time, most statues were removed in response to San Francisco's earthquake. Because of deterioration, the remaining statuary was taken down in 1948. New statuary was created for the restored building and returned to the west, north, and south sides *(Top)*. Five original statues have remained in the west pediment *(Bottom):* Minerva is the central figure; to her left sit Justice and Mining; to her right sit Education and Industry. All original statuary had been made by Italian sculptor Pietro Mezzara.

The Capitol's cornerstone was first laid on May 15, 1861. In the following December and January, floods forced builders to raise the first floor one story. Dirt was brought in to raise the ground and the cornerstone disappeared. It was located in 1952 but not removed until 1978 when uncovered during the restoration. The copper casket inside the cornerstone contained old coins, newspapers, documents, and the original rendering of the Capitol's conceptualized design *(Top)*.

The cornerstone was relaid on May 18, 1978 *(Bottom)*. It now contains some original objects as well as new ones: a photograph of the workers on the dome, a program from the Capitol Cornerstone Relaying Ceremony, a copy of Assembly Bill 2071 which provided for the restoration, the seismic study report on the Capitol, messages from the legislators, coins, stamps, a Bible, the 1978 Sears Catalog, and two bottles of wine from one of the oldest wineries in California.

INTERIOR RES-
TORATION. *(Bottom right)* Cast iron
decoration was stripped from the
dome and bright Victorian colors
whitewashed as attempts to mo-
dernize the Capitol began. Nothing of
the original design remained for restoration workers except a photograph from 1890 *(Top left).*

The painter's sketch displayed
the colors to decorate the dome and
rotunda *(Top right).*

The hot light from a xenon flash lamp removed old paint in layers. Some original colors were found, as well as ghost images of the earlier decoration *(Top left)*.

This stylized fleur-de-lis pattern *(Bottom left)* was painted on canvas in a studio prior to application in the dome.

Most decorative gold was created with a brass alloy called Dutch metal. A slow drying glue was applied to the surface *(Top right)* before the thin sheets of metal were brushed on *(Bottom right)*.

25

Some of the decorative pieces were sketched on plywood which matched the dome's curvature *(This page; top left)*, then modeled in clay *(Top center)*. A mold was made from the clay piece so plaster castings could be created *(Top right)*. These plaster festoons were highlighted with Dutch metal *(Bottom)*.

(Opposite page) The dome and rotunda once again display their original Victorian beauty.

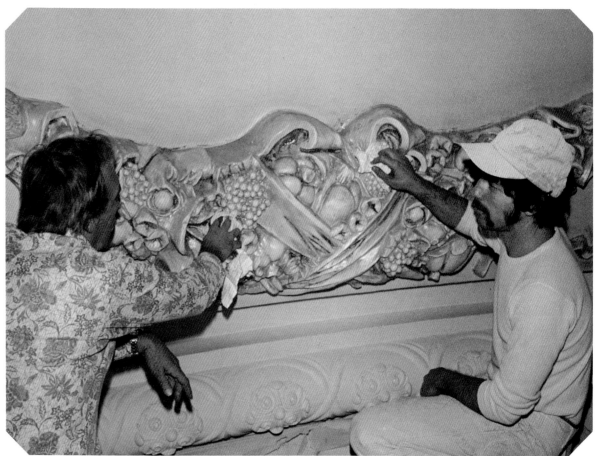

In 1913, rotunda niches on the first floor were bricked over and the artwork covered to display twelve canvas murals. A historic photograph *(Bottom)* revealed the original rotunda interior.

(Top left) Arthur F. Mathews, a San Francisco artist, was selected to paint the murals which would replace the original rotunda decoration. The murals depicted historical epochs of California's history and represented four themes: the coming of the "White Gods"; the Spanish and Mexican occupation; the Anglo-American occupation; and the achievements of civilization in California. The murals can be viewed in the Capitol's basement rotunda *(Top right)*.

When the murals were removed during the restoration, the original niches were still in place *(Top left)*. These "marbleized" or painted originals served as models for reproduction. The new, hand-painted niches *(Bottom left)* illustrate the revival of a nearly-lost art.

Use of the xenon lamp revealed traces of the original decoration on the rotunda walls *(Top right)*. This artwork was painted on canvas and contains the griffin, a mythological animal with the head of a lion and the body of an eagle.

The original marble mosaic on the second floor did not need to be reproduced. However, the mosaic presented a difficult problem: how could it be saved when the walls and floors were removed? The floor was photographed from above to retain an exact record of the design. Then, using a special saw for concrete, the floor was cut into squares *(Top right)*. Plywood was glued to the marble surface *(Bottom right)* so the squares could be lifted out *(Bottom left)*.

The marble mosaic consists of approximately 600,000 pieces. Each piece was cleaned, polished, grouted and glued on kraft paper. The squares were stored in boxes and later reinstalled in sheets.

The restored 1906 mosaic *(Bottom)* shows the Golden Poppy, California's state flower.

The original Eureka tile was installed in the Capitol's first floor corridors around 1896. Due to excessive wear of the original, reproduction tile was created for the restored building.

The first step in reproduction was to spray a base coat glaze over the entire ceramic tile *(This page; top left)*. To create the designs, stainless steel screens were used, one screen per color. Each color was a different engobe, or liquid glaze. Over five hundred screens were required to recreate the Minerva mural.

(Opposite page) The image of Minerva, the Roman goddess of wisdom, appears frequently in the Capitol. Early Californians believed Minerva's birth in mythology was similar to the way in which California became a state. Neither Minerva nor California had a childhood. Minerva emerged fully grown from the head of Jupiter, her father. California emerged into statehood without first becoming a territory.

"Eureka" is California's state motto and in Greek means "I have found it." In 1848, James Marshall discovered gold at Sutter's sawmill, and the rush to California began.

The original Minton-Maw quarry tile on the second floor was manufactured in England. Since most of the tile was in bad condition, reproductions were made from standard United States quarry tile. Original tile that could be saved was placed on the stair landing; new tile can be seen on the rotunda floor.

The original Belgium black and Vermont white tile in the first floor rotunda was badly worn. Recreated rotunda and portico tile came from Belgium and Alabama. Original tile remained on the first floor foyer.

(Top) The statue of Christopher Columbus and Queen Isabella at the Court of Spain has decorated the first floor rotunda since 1883. The statue was a gift to the Capitol from D. O. Mills, a well-known banker of the time. The marble statue was created by an American sculptor, Larkin Goldsmith Meade, in Italy.

Multicolored parget plaster work was once common in the Capitol, but modernization led to either whitewashing the Victorian colors or removing the work entirely.

A full-color fragment of this poppy design *(This page; top left)* was discovered when workers removed an old ventilation duct. The frieze and ceiling in the room had been painted white except where the duct accidently preserved the colors.

A full-scale, perforated drawing *(Top right)* was made of the frieze and ceiling. The design was transferred by applying charcoal powder over the perforations onto the surface.

Artists used a pastry tube and other tools to apply a plaster mixture directly on the wall and ceiling where it was sculpted into the final forms *(Bottom)*.

Finally, the parget work was painted to resemble the original colors *(Opposite page; top left and right)*.

This parget frieze and ceiling can be viewed in the Capitol's Archive Exhibit Room located on the first floor *(Lower right)*.

Other types of decorative plaster work include "run-in-place-plaster," which was created at the top of the walls using a template *(Top left)*. Corners were mitered by hand to finish the work *(Top center)*. Another type of plaster, staff plaster, was cast into latex molds *(Top right)* then adhered in place *(Bottom left and bottom right)* with wet plaster and mechanical fasteners.

All of the Capitol's ceilings and floors were originally constructed with a brick arch system. This system was recreated as a model in the basement dining area *(Bottom)* to show visitors the original design.

The monumental staircases *(Top left)* were removed in 1906 to allow for additional office space. A limestone and marble facade *(Top right)* took the place of the west stairs and led into the new offices.

Several sources were used to re-create the staircases: a historic photograph, the carpenter's original layout block and curved wood backing *(Bottom left)* which had been uncovered when the facade was removed, and one original newel post and balustrade section found in a Sacramento church.

Honduras Mahogany was used to re-create the staircases *(Bottom center and bottom right)*; originals had been made of walnut, mahogany, and redwood.

The reproduction newel post light fixtures *(Left)* resemble the gas fixtures which once lit the halls.

(Top right and center right) Both legislative chambers had evolved to modern spaces by the 1970s. Fluorescent lights replaced crystal chandeliers, stencil work was painted over, and marbleized columns stood where fluted columns had once been.

When the Capitol first opened in 1869, a *Sacramento Daily Union* reporter found the early chambers to be perfect. "This happy mingling of colors by the painter's brush, this ingenious carving by the skillful worker in wood, that horn of plenty, all tend to impress the mind with pleasurable and patriotic emotions."

(Top left) The Senate Chamber, circa 1890. *(Bottom)* The 1900 Assembly in session.

Historic photographs showed that the early chambers had two rows of columns and pilasters. Since the originals had been removed, recreations were made. Columns were cast in iron based on the pattern shown in the historic photograph.

(This page; top left and bottom left) The coffered ceilings were precast in plaster, delivered in large sections, and hung in place in the chambers.

Fragments of earlier decoration *(Top right and center right)* which had been removed in the 1906 remodeling were found underneath the floor in the Assembly Chamber when the building was gutted. These pieces aided craftspeople in recreating the ceiling pendants and other decorations.

Stencil and border designs in the Senate Chamber were hand-painted in place *(Opposite page; top left and bottom left)*.

The ceiling in the Assembly Chamber displays various California wildflowers *(Right)*. The flowers were painted in a studio on canvas, then applied inside the coffers.

(Top left) Statues of the Roman goddess Minerva once overlooked both chambers. Today, Minerva appears only in the Senate *(Top right)*. The artist recreated the statue in clay. A mold was made from the clay form so a plaster figure could be created *(Bottom)*.

The ceilings in both chambers appear approximately eight feet lower than they were in 1900. During the 1906 remodeling, the chambers were gutted and the space lowered to accommodate the fourth floor addition.

Reproduction chandeliers were made of European lead crystal *(Left and top right)*. The original gas fixtures could be lowered to the floor for lighting.

The Senate Chamber clock entablature *(Center right)* was recreated to resemble the original. An adapted version appears in the Assembly Chamber *(Bottom right)*.

(Following page) The forty senators and eighty assembly members use restored original desks. The desks and front daises were manufactured through the John Breuner Company and were in place when the 1869 legislature first occupied the Capitol.

The members of the Assembly are elected for two-year terms and each district consists of approximately 300,000 people. The Assembly Chamber is equipped with an electronic voting system; members press a button on their desk to indicate their vote. Each vote and a tally appears on the panel in the front of the chamber.

Gold leaf highlights the Latin phrases in the chambers and shows the motto of each house. "Legislatorum Est Justas Leges Condere" appears in the Assembly and means "It is the Legislator's duty to establish just laws."

(Preceding page) Senate members are elected for four-year terms. Each senator represents approximately 600,000 people. Instead of an electronic system, senators use a more traditional roll call vote. The motto in the Senate reads, "Senatoris Est Civitatis Libertatem Tueri" — "It is the Senator's duty to guard the liberty of the state."

(This page, top) Today, modern functions of government are carried out in historic settings, as this committee hearing room illustrates.

The more modern fourth floor *(Bottom)* contains working office space.

(Top left) Some of the original elevator grille work was found under the existing modern facades. Reproductions were made to resemble the 1906 passenger elevators (Bottom left).

Two stained glass state seals (Right) were installed on the second floor chamber entrances during the 1906 remodeling.

HE MUSEUM.

Seven turn-of-the-century house museum rooms and two exhibit rooms occupy the west side of the first floor. These rooms provide an insight into another era. The rooms were once real offices, animated by real people. The office workers stored leather ledger books in the cabinets, heated tea water in a silver kettle on the coal-burning stove, and washed black ink from their hands in the corner lavatory.

What is a house museum? It is the accurate placement of objects in a setting to tell a story about the past. The Capitol's museum shows California's Executive Branch of government at the turn-of-the-century.

(Top left) Treasurer Truman Reeves' 1898 office and the office recreated as a museum *(Top right)*. Historically, the office served as a bank for state workers.

(Center right) Governor Pardee's 1906 Main Office. In the museum *(Bottom right),* the documents and letters reflect the feeling of crisis that immediately followed the San Francisco earthquake.

How did the museum staff begin the task of recreating the historic offices? Brainstorming sessions produced a system: historic photographs, receipts and documents had to be located; the time period was researched and specific objects identified through old catalogs; the work was divided among the staff. Approximately 10,000 objects were needed to create the museum. The objects were stored in a warehouse until installation in the Capitol.

Other museum rooms include the Governor's Anteroom and Private Office, the 1933 Treasurer's Office, and the Attorney General's Office. The State Library and Archive Exhibit Room display original historic documents.

(Top left) Secretary of State Charles F. Curry's office was crowded in 1902 because of his varied responsibilities. The museum *(Bottom)* recreates this feeling.

Creative Direction
  Pat Davis Design, Sacramento

Design
  Brenda Walton Design, Sacramento

Calligraphy
  Brenda Walton

Typography
  Garamond by ATG/Ad Type
  Graphics, Sacramento

Printing
  Graphic Center, Sacramento

Paper
  Quintessence

Photography
  California State Capitol Museum:
  Page 5; 24 (Top right); 42 (Bottom); 52
  (Top left, center right); 53 (Top left).

  California State Capitol Restoration
  Project: Michael H. Casey, Page 46 (Bottom left). Robert Dunham, Page 46
  (Bottom center). Dale E. Dwyer, Page 4
  (Bottom); 7 (Bottom right); 8; 9; 10; 11;
  12; 13; 14; 15; 16 (Right); 17; 18; 19
  (Top); 20; 21; 22 (Top left, top center);
  23 (Bottom); 24 (Bottom); 25; 26; 29
  (Top left, right); 31 (Top); 32 (Bottom);
  34; 35 (Bottom); 36; 37 (Bottom left);
  38; 39 (Top); 40 (Top right, bottom
  right, bottom center); 42 (Top right,
  center right); 43; 44; 45 (Left); 51 (Top
  left). Lynn G. Marlowe, Page 22 (Top
  right); 37 (Top right); 46 (Bottom
  right); 47 (Top right, center right); 53
  (Top right). Wendy Welles, Page 32
  (Top left, top right).

  California State Department of Parks
  and Recreation: Page 31 (Bottom); 37
  (Bottom right, top left); 45 (Right); 52
  (Top right, bottom right).

  California State Department of
  Transportation: Page 3 (Center left, bottom left); 6; 28 (Top left); 53 (Bottom).

  California State Library: Page 2 (Top
  left); 3 (Top left, right); 4 (Top); 28 (Bottom); 40 (Top left); 42 (Top left); 46
  (Top left).

  Collection of the Sacramento
  Museum and History Divison: Inset
  opposite title page; Page 24 (Top left).

  Robert A. Eplett: Back cover.

  Higgins Agricultural Library, University of California, Davis: Page 2 (Bottom).

  James Mazzuchi Photography,
  Placerville : Page 30

  John Palmer: Page 22 (Bottom); 52
  (Bottom left).

  The Lester Silva Collection: Page 2
  (Top right).

  Bob Van Noy: Front cover; opposite
  title page; opposite table of contents;
  opposite introduction; Page 19 (Bottom); 27; 28 (Top right); 29 (Bottom
  left); 33; 35 (Top); 39 (Bottom); 41
  (Left); 46 (Top right); 47 (Bottom right);
  48; 49; 50; 51 (Right, bottom left).